Philipp Gimpel

The impact of the global financial crisis on global cities

Shifting importance of leading global cities?

GRIN Verlag

Bibliografische Information der Deutschen Nationalbibliothek:

Die Deutsche Bibliothek verzeichnet diese Publikation in der Deutschen National-bibliografie; detaillierte bibliografische Daten sind im Internet über http://dnb.d-nb.de/ abrufbar.

Impressum:

Copyright © 2013 GRIN Verlag GmbH
Druck und Bindung: Books on Demand GmbH, Norderstedt Germany
ISBN: 978-3-656-52616-2

Dieses Buch bei GRIN:

http://www.grin.com/de/e-book/263318/the-impact-of-the-global-financial-crisis-on-global-cities

GRIN - Your knowledge has value

Der GRIN Verlag publiziert seit 1998 wissenschaftliche Arbeiten von Studenten, Hochschullehrern und anderen Akademikern als eBook und gedrucktes Buch. Die Verlagswebsite www.grin.com ist die ideale Plattform zur Veröffentlichung von Hausarbeiten, Abschlussarbeiten, wissenschaftlichen Aufsätzen, Dissertationen und Fachbüchern.

Besuchen Sie uns im Internet:

http://www.grin.com/

http://www.facebook.com/grincom

http://www.twitter.com/grin_com

UNIVERSITÄT HAMBURG

Fakultät für Mathematik, Informatik und Naturwissenschaften

Seminar paper

The impact of the global financial crisis on global cities – shifting importance of leading global cities?

Author: Philipp Gimpel

1. Semester Geography (Master)

Advanced seminar: "Spatial dimensions of the global crisis"

Winterterm 2012/2013

Seminar facilitator: Prof. Dr. Jürgen Oßenbrügge

Table of Content

1. Introduction

In times of intensifying globalization processes and interweaving of the global economy, it is very interesting to find out, what are the actors or places where these processes are controlled. Are there any important places which have key functions to command and control? Many scientists are gone further into this question, and many concepts appeared to find out these centres of commanding and controlling. The most popular concept is the global city concept by Saskia Sassen. But for the most part, all concepts assume that there must exist some centres or cities, which include functions to control the world economy. And if there are such centres, are there any centres which are more important than others and are there any forms of hierarchies? Thus are there any processes which lead to a shift within the hierarchies like the global financial crisis?

This paper seeks to answer these questions, especially if there is a shift of the leadership of global cities from predominant centres before and after the crisis.

2. Global Cities – Key Concepts

The intensifying processes of globalization in economic, labour organisation and communication fields, necessitates centralized functions of management. A transnational bundling of these fields requires a practical place of location, the so-called "global city", or as Bronger quotes Feagin & Smith:

„World Cities are the anchorage points which keep the capitalistic world economy together." (Feagin & Smith, in Bronger, 2011)

The „global city"-term is not a new one especially when associated with the term „world city", because both terms are used similarly. So it is not astonishing that there is a variety of definitions. (view table in Bronger, 2011: 332 f.)

To sum up the criteria for a "global city" shown in this table (view table in Bronger, 2011: 332 f.) one can say that "global cities" are centres of politics, economics, communication, culture and science. Furthermore these cities are important anchorage points for traffic, transport and trading (e.g. "global cities" are places for important harbours and international airports), but the emphasis lies on economics. With headquarters of transnational corporations they are control centres within the organisation of the global economy. As enmeshed in the global economy, transnational economical acting in these cities has an impact on the global economy. Another aspect is the population size. "Global cities" cannot be defined just by the population growth but some say that one has to comprise the population size like Kiuchi (1959), Birkenhauer (1982), Heineberg (1989) or Juchelka (1996). In contrast, "Megacities" for example can just defined by population size. At a glance, "global cities" are international centres and hubs of a transnational organised and capitalistic economy, cultural and intellectual mediators between the continents – in contrast to megacities which act as national centres."

2.1 Key Concept – John Friedmann's "World City"

The concept of the world city goes back to two texts written by John Friedmann and Goetz Wolff in 1982 and by Friedmann in 1986. They elaborate "the rise of a global urban network in the context of a major geographical transformation of the capitalist world-economy" (Derudder et al., 2012: 74). As already mentioned above and acknowledged by Friedmann, this transformation demands for "a number of command posts in order to function" (Derudder et al., 2012: 74) and in the world city concept, world cities are presumed to be the "geographical emanation of these command posts" (Derudder et al., 2012: 74). As perhaps expected, Central Business Districts are not the "territorial basis" (Derudder et al., 2012: 74) for the world city. In terms of the world city concept, a city is a "spatially integrated economic and social system" in the space of a pre-existing location. (Derudder et al., 2012: 74)

Friedmanns work was built upon Frobel et al.'s New International Division of Labour (1980) and the "role of international financial centres on global space economy" (1981). (LEE et al., 2012: 335) But Friedmann's work is principally based on Wallerstein's analysis on world-systems. In his work, Wallerstein emanates from a capitalistic system that contains "hierarchical and spatial inequality of distribution" (Derudder et al., 2012: 74). The inequalities are results of a concentration of monopolized and high-profit production in core areas. And the spatial inequality is distinguished by the division of labour which reflects in the cityscape as a "tripolar system consisting of core, semi-peripheral and peripheral zones" (Derudder et al., 2012: 74).

The intention to frame the world city concept was to prescind "from the role of territorial states in the reproduction of this spatial inequality" (Derudder et al., 2012: 74). In addition, as Mann (1986) and Dogshon (1998) mentioned, it comes along that the world-economy is not territorially managed, but more radially. Economic and political power of core territories is arranged by "well-defined routeways" which connect control centres.

Here, core territories do not mean territorial states but a hierarchy of centres like world cities. And as Taylor notes, the modern world-system is defined by its networks and world cities are the nodes in such networks of power and dominance. Some other forms of command and control are performed in world cities, like geopolitical and ideological control. But for all explanations, one has to consider the most important variable, namely the economy. Thereby, the focus lies on "corporate headquarters and international financial institutions and agencies" (Derudder et al., 2012: 75).

2.2 Key Concept – Saskia Sassen's "Global City"

The global city concept is well known and was developed by Saskia Sassen As it sounds from the world city concept, centrality plays also an important role in the global city concept. The focus is on the "functional centrality in the global economy" (Derudder et al., 2012: 75), whereat here the focus is on the "attraction of producer service firms to major cities that offer knowledge-rich and technology-enabled environments" (Derudder et al., 2012: 75). Producer service complexes emerged in the 1980s and 1990s as such firms moved afterwards their global clients. These complexes stand at the roots of the global city-formation. This contains "a shift of attention to the advanced servicing of worldwide production" (Derudder et al., 2012: 75). So, the focus shifts to the exercising of global control. The essential for organizing the world economy is power. According to Sassen, producer service firms with their "transnational, city-centred spatial strategies" (Derudder et al., 2012: 75) created a network with installed offices in "major cities in most or all world regions" (Derudder et al., 2012: 75). With this network of offices, there are connections between the service complexes of the cities. There is a fading to a "formation of transnational urban systems" and this affords "a new geography of centrality" (Derudder et al., 2012: 75). The functional centrality of these global cities becomes more and more disconnected from their proximate surroundings or their national economies. Here, Sassen decides to study the "core dynamics" (Derudder et al., 2012: 76) and less the city as the built environment.

	World Cities	Global Cities
Key author	Friedmann	Sassen
Function	Powerhouse	Centre for servicing of global capital
Key agents	Multinational corporations	Producer service firms
Structure of the network	Reproduces (tripolar) spatial inequality in the capitalis world system	New geography of centrality and marginality cutting across existing core/periphery patterns
Territorial basis	Metropolitan region	Traditional CBD or a grid of intense business activity

Tab.1: Principles of Friedmann's and Sassen's concepts, source: Derudder et al., 2012: 76

Tab.1 gives an overview of the key principles of the world city concept and the global city concept. Friedmann focuses on the whole metropolitan region, Sassen on the traditional Central Business District. Friedmann considers these kind of cities as "centres of dominance and power" (Derudder et al., 2012: 76) and Sassen as "production centres for the inputs that constitute the capability for global control" (Derudder et al., 2012: 76). So, there arise different perspectives on the cities as anchorage points of transnational networks relating to the cities functions, their key agents, the structure of the networks and the territorial bases. (Derudder et al., 2012: 76)

3. Methodology – The interlocking network model

When we speak about global cities, we also speak about these cities under globalization dynamics. And if we think about a network where global cities are connected with each other, then the idea of relations between these cities within a network arises. To analyse these relations between cities we need a method. But what method could be used to explain or illustrate relations between cities within a network and what's important, what value to differentiate the degree of relationship within the network?

There are about 50 types to illustrate inter-city relations. Peter J. Taylor introduces his interlocking network model or world city analysis. In his analysis which derives from Jacobs' (1969) ideas he, like Jacobs did, comes from the idea that "cities come in groups and need each other" (Derudder et al., 2012: 53). That means that network relations are horizontal rather than vertical. Taylors' theoretical background derives from Sassen's global city concept. According to Sassen, many service firms followed their clients. The outcome of this was that they built up and invested in a network office around the globe. And "it is the work done in these office networks that defines today's world city network" (Derudder et al., 2012: 53). That means that with the everyday work in these offices, the cities are economically connected. Furthermore, the development of telematics and "inter-personal links 'interlock' cities through advanced producer servicing" (Derudder et al., 2012: 53). But to measure flows "of inter-city business interlocking" (Derudder et al., 2012: 53) with appropriate data is not possible. According to Taylor, the only solution is an indirect measuring of these flows (Derudder et al., 2012: 53).

Taylor developed the interlocking model with three formulas. First, one measures the nodal size of a city a. Here, S_a means the nodal size.

$$S_a = \sum_i v_{ia}$$

There are n service firms given which have offices in m cities and there is the service value v_{ij} where i stands for the activity of a firm and j for the city where this activity takes place. Within this formula, one measures "the quantity of flows between two cities generated by a firm" (Derudder et al., 2012: 54). To define the network activity of a city or rather to sum up all inter-city products for all firms in one city with all other cities inside the network, one needs the following formula:

$$C_a = \sum_j \sum_i v_{ia} \cdot v_{ij}$$

Here, C_a stands for the network connectivity of city a. And as already mentioned above, "cities come in groups and need each other" (Derudder et al. 2012: 53). So, here we can measure the relation between city a to all cities in the network. But to facilitate comparisons, Taylor developed a third formula where network connectivity turns out to be "proportions of the highest connectivity recorded" (Derudder et al., 2012: 54):

$$P_a = C_a/C_h$$

Thereby, C_h means the highest connectivity measure and P_a stands for the city a's proportion of C_h. (Derudder et al., 2012: 54)

4. The global financial crisis

Before we can understand the impacts of the global financial crisis, we need to consider what circumstances enabled this crisis. There is a general agreement that the period to the 1970s was characterised by a relatively solid Fordism. There we had a standardised mass production for the requirement of the mass with pegged wages on development of productivity and prices. These wages were mediated between major enterprises, unions and the Keynesian state. This model was assumed by strong controlled financial markets and restriction for transnational capital flows. Since the end of the 1960s this model was increasingly pressurized and debilitated by the breakdown of Bretton Woods (Nölke 2009: 124-125).

4.1 Financialization

The development of capitalism since the late 1970s is actually analysed by economists among the term "financialization". That is the process where increasing shares of assets and earnings of households are made out of financial activities in contrast with the shares made out of production or hired labour. So, we can observe a shift importance between and within corporations where the financial sector gets increasingly significant. In addition, the impact of fluctuations and crises are more eminent on the productive sector because of a stronger linkage between corporations and the financial sector. That is also possible because the financial sector takes on a life of its own to a greater extent and has got a bigger dominance over the economy. (Nölke, 2009: 125)

Over the years, not only profitability of the financial sector developed much better than the productive sectors but also the share of profits made out of financial market activities within corporations of the productive sectors have increased. Some of them have even more assets because of financial transactions than with their primarily production. The increasing profits are based on a great number of financial innovations like derivatives and securitizations what was especially encouraged by investment banks. First, financing of corporations was to the fore, but financing has expanded to other sectors like consumer goods, student loans and especially real estate. Notably in the USA, mortgages were securitized; non-payment risks were displaced to purchaser of these securitizations. The expansion of the real estate financing market proceeded what could be seen as the precondition of the subprime crisis. (Nölke, 2009: 126-127)

Financialization is not limited to particular corporate finance but embraces also the "mass investment culture" (Nölke, 2009: 127) of households, so that they increasingly depend on financial markets, especially institutional investors, in terms of pension funding. (Nölke: 127-128)

Whereas, financialization depends on technological factors like telecommunication and financial mathematics and supported by demographic developments it is not just a structural development of capitalism, but also a political promoted project. The essential cause for a transition to financialization is the liberalisation of the international movement of capital at the end of Bretton Woods 1973 and the subsequent measures for the reformation of national capital markets. Financialization gained momentum in the era of Alan Greenspan at the top of the Federal Reserve of the United States as he revised the policy of high interest rates, what facilitates to borrow credits for capital market speculations. Further it gains momentum because there is an attempt for seeking suitable investment opportunities for accumulated surpluses. With these saving gluts, accumulation of assets and demand shortfall by social segmentation are involved. Besides stated points, the Thatcher- and Reagan-Administration, financialization as part of neoliberal policy and the Glas-Steagall Act under the Clinton-Administration (less segregation of investment banks and commercial banks) encouraged financialization. (Nölke, 2009: 128)

4.2 Finacialization and the global financial crisis

Finally, it is not clarified in what extent financialization has to do with the outburst of the global financial crisis. But one assumes that there is no actual global financial crisis without financialization and that a crisis with this extent could never have broken out under Fordism with its highly regulated markets. One cannot observe bigger crises between the 1930s and the 1960s, but crises became more frequent, e.g. "start-up crisis", "Russia-crisis" or "East Asia-crisis", so that some observers speak about a crisis of capitalism of financialization. (Nölke, 2009: 130)

In some ways, the financial crisis stands in a direct linkage with financialization, especially with a product of financialization. There is the transformation of conventional mortgages saved by a counter-value, to anonymised and securitised financial products. Traditional mortgages are not directly affected by the crisis. Financialized economies are more impacted by booms and following crises than "conventional" economies because shares and financial products are more suitable for speculations than corporations and real estates. Furthermore, phases of the so-called "boom and bust" are a necessary part of financialization, that means that to provide financialized economic growth, new investment perspectives are required again and again. (Nölke, 2009: 130)

Basically, one can say that the phase of financialisation of capitalism is more crisis-prone than Fordism. So, some experts even expected the actual global financial crisis. And factors, like political arrangements which served the intensification of financialization, intensified the crisis. Nearby the arrangements of liberalising the financial market, there is a switch of accounting standards of historical acquisition values to the actual market valuation, what implies increasing market dependence. This switch is induced by the transition to financialization and contributes further financialization. But within the crisis management, this switch is partially corrected. While the valuation of financial goods to market prices in a boom phase leads to a higher valuation of these goods (and an increasing share of financial goods on assets), the same valuation in a decline stage leads to a lower valuation of these goods and an intensification of the crisis of banks and other actors of the financial market, because they are compelled to sales. That again leads to a price collapse. Even this coherence was known before the crisis. (Nölke, 2009: 131)

To understand finally the dimensions of the crisis, we need to regard the immense growth of financial markets. Without any financial markets at a "hypertrophic" scale the subprime-crises could have never lead to the strongest economic crisis since the 1920s. Not least, the crisis had a lesser impact on continental Europe than the higher financialized economies in the Anglo-Saxon heartland. (Nölke, 2009: 132)

Nonetheless, besides these high financialized economies like the USA or Great Britain (with the upgrading of the financial sector and the simultaneous erosion of the industry), which are strongly affected by the crisis, even the economies of continental Europe and Japan are less strong affected by the crisis. (Nölke, 2009: 132)

In the USA and Great Britain the crisis got around intervening by communization and financial support by the public sector, because of bank collapses (view Tab.2). There were also spectacular collapses on continental Europe (view Tab.2). (Nölke, 2009: 132)

Emergent economies like India, South Africa or the states Latin America have been relatively less affected by the crisis. They are more affected by the following recession of above-mentioned countries as customers of goods and resources. Also China and Near Eastern petroleum exporting countries have been affected, strongly affected in some cases, if they have used Anglo-Saxon banks for purchasing and suffered losses. (Nölke, 2009: 134)

	Country	Corporation	Total allowances and losses since January 2007 (billion US $)
1	US	Citigroup	60.8
2	US	Wachovia	52.7
3	US	Merill Lynch	52.2
4	US	Washington Mutual	45.6
5	CH	UBS	44.2
6	UK	HSBC	27.4
7	US	Bank of America	21.2
8	US	JPMorgan Chase	18.8
9	US	Morgan Stanley	15.7
10	GER	IKB Deutsche Industriebank	14.3
11	UK	Royal Bank of Scotland	13.8
12	US	Lehman Brothers	13.8
13	GER	Deutsche Bank	10.1
14	CH	Crédit Suisse	10.1
15	US	Wells Fargo	10
16	FRA	Crédit Agricole	8.6
17	UK	Barclays	7.5
18	CAN	Canadian Imperial Bank of Commerce	7.1
19	BENELUX	Fortis	6.9
20	GER	Bayerische Landesbank	6.7
21	UK	HBOS	6.6
22	NL	ING	6.5
23	FRA	Société Générale	6.4
24	JAP	Mizuho Financial Group	6.1
	Worldwide		*586.2*

Tab.2: Losses per bank owing to the crisis, source: Nölke, 2009: 133

In Tab.2 one could confirm Nölke's statements that the Anglo-Saxon countries have most suffered from the crisis. There are eight out of the top ten corporations who have their headquarters in the Anglo-Saxon space and fourteen out of all twenty-four. And even the statement that also countries on continental Europe and Japan have suffered from the crisis. There are two out of the top ten and nine out of twenty-four corporations that have their headquarters in continental Europe and one that has its headquarter in Japan.

When we speak just about the financial sector or financial markets, then we could endow Nölke's statements but as he always talks about whole countries we cannot support his statements. And if he would speak about the financial sectors, he also should speak about real estate and insurance companies, which are missing in Tab.2. But in the following, one can use this table as an indicator for further investigations.

5. The crisis and its impacts on global cities

5.1 Global cities in the global space-economy

As there has been the global financial crisis and a collapse of some banks and a disruption of the global financial system, there also must have been an impact of the crisis on global cities or the command and control centres of the global economy. If we assume, as Taylor and others did, that these financial centres and global cities are entangled within a global encompassing network and that these cities are related to each other, than we have to reckon that there is an impact on these cities. But the question is in what extent these cities were impacted by the global financial crisis.

This raises the question what methodology we should use to get useful results and what is more important, to get comparable results. First, there is the interlocking network model developed by Taylor. But this is just partly useful, because it examines the network connectivity of cities. If we want to know if there has been a shifting of leading global cities (or as Taylor et al. 2009a called it a geo-economic transition), we need to create a hierarchy because the term "leading" implicates a hierarchy.

For that, Taylor et al. developed indices to measure the importance of financial centres or what they call it "command-and control-cities in the global-space economy" (Taylor et al., 2009a).

First, there is the financial command index (FCI) where each city with "a top-five firm headquarters" (Taylor et al., 2009a: 8) gets a score of 10. Cities with headquarters ranked with 6-10 get a score of 8, cities with headquarters 11-20 get a score of 6, with 21-30 get a score of 5, with 31-40 get a score of 4, with 41-55 get a score of 3 and with 56-75 get a score of 2. The composite data according to Forbes includes "rankings for sales, profits, assets, and market value". (Taylor et al., 2009a: 8)

And second, there is the business command index (BCI) where "each city with a top-fifty firm headquarters" (Taylor et al p.10) gets a score of 12, cities with headquarters 51-100 get a score of 11, with 101-200 get a score of 10, with 201-300 get a score of 9, with 301-400 get a score of 8, with a 401-500 get a score of 7, with 501-600 get a score of 6, with 601-700 get a score of 5, with 701-800 get a score of 4, with 801-1200 get a score of 3, with 1201-1600 get a score of 2 and with 1601-2000 get a score of 1. The data, also according to Forbes was computed based on "the top 2000 firms in the world". (Taylor et al., 2009a: 10)

Taylor et al. computed both indices and the results are evident from Tab.3 and Tab.4. Two cartographic figures illustrate the spatial distribution of financial command-and-control centres and of business command-and-control centres.

Rank	City	FCI
1	New York	100.00
2	London	60.71
3	Zurich	37.50
4	Paris	35.71
5	Toronto	30.36
6	Tokyo	28.57
7	Charlotte	26.79
8	Edinburgh	25.00
9	Amsterdam	23.21
10	Beijing	23.21
11	Brussels	23.21
12	Munich	19.64
13	Washington	17.86
14	Basel	14.29
15	Frankfurt	10.71
16	Minneapolis	10.71
17	Omaha	10.71
18	San Francisco	10.71
19	St Petersburg	10.71
20	Melbourne	8.93

Tab.3: Top twenty cities according to the FCI in 2008, source: (Taylor et al., 2009a: 8)

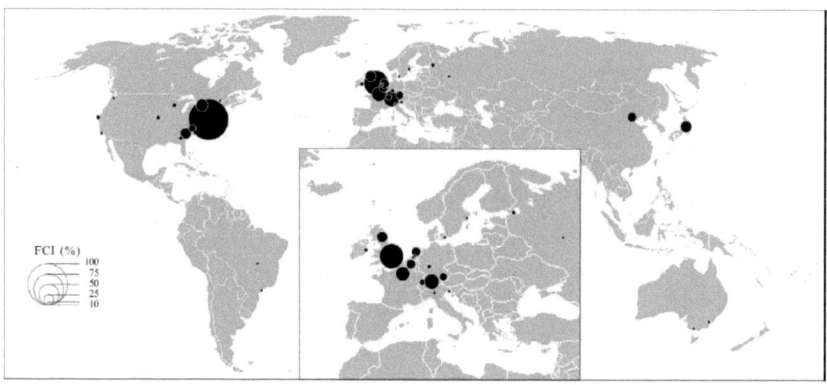

Fig.1: Financial command-and-control centres in 2008, source: (Taylor et al., 2009a: 9)

Rank	City	BCI
1	Tokyo	100.00
2	New York	70.94
3	London	68.49
4	Paris	53.96
5	Houston	25.47
6	Seoul	23.31
7	Chicago	21.44
8	Osaka	20.00
9	Beijing	19.42
10	Madrid	19.14
11	Stockholm	18.71
12	Los Angeles	18.13
13	Toronto	17.84
14	San Jose (CA)	17.70
15	Washington	16.40
16	Hong Kong	16.26
17	Sydney	12.81
18	Dallas	12.66
19	Taipei	11.65
20	Melbourne	11.37

Tab.4: Top twenty cities according to the BCI in 2008, source: (Taylor et al., 2009a: 10)

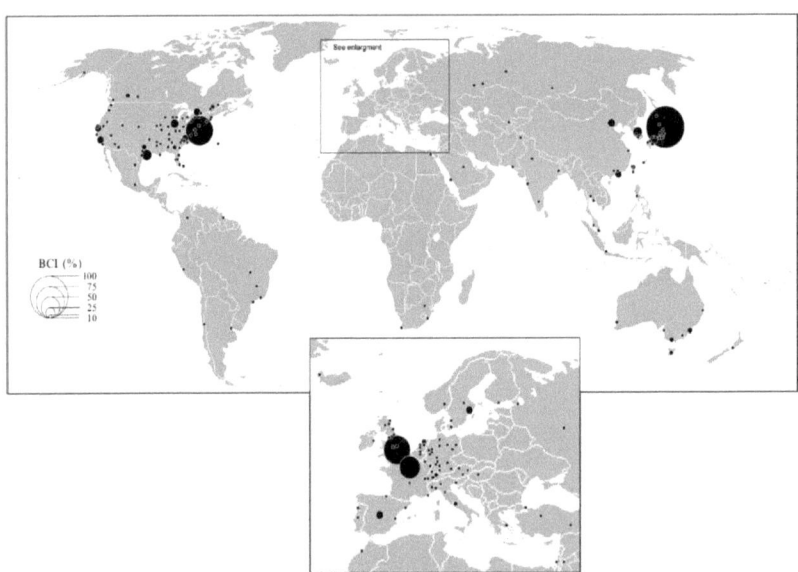

Fig.2: Business command-and-control centres in 2008, source: (Taylor et al., 2009a: 11)

All scores are summed up for each city so that the results of Tab.3 and Tab.4 come to existence. What attracts attention is that for the year 2008, cities computed with the FDI, nineteen out of twenty cities are based on the northern hemisphere and that nineteen out of twenty cities are based in North America and Europe. But what Taylor et al. 2009a state is more distinct in terms of the concentration of command functions of these cities, that twenty-nine out of thirty-five are based in North America and Europe. This can be seen in Fig.1. Additionally we see the dominance of New York and then only London with more than a half of the FDI of New York.

For the BCI, many more data was analysed, so the results shown in Fig.2 are more detailed or more widespread. In Tab.4 it stands out that Tokyo was the leading business command-and-control centre, followed by New York, London and Paris. All of them are the dominant leading cities according to the BCI. Then there are some unexpected cities like Houston or San Jose (CA). Taylor et al. explain the fifth place of Houston "as the world's oil and gas control-and-command centre". (Taylor et al., 2009a: 10)

The table ranked by the BCI includes multiple cities of the USA, Japan and China. The residual cities "are leading cities of medium-sized national economies". (Taylor et al., 2009a: 10) Taylor et al. state that one can already recognize some winning and losing cities. But the comparison with other years is missing, so that it is hard to subscribe the statements of Taylor et al. that there is an ongoing transition. They argue that Tokyo is the main evidence and also other cities could be indicative that a transition is ongoing. (Taylor et al., 2009a: 8-10)

"The relative decline of Tokyo's banks since 1990 is shown by the city's 6th ranking. Even at this early stage in the transition we can note some losers and winners: Charlotte and Edinburgh are, perhaps, surprise top-ten entries but they are unlikely to retain this preeminence: Wachovia's demise will hit Charlotte and Edinburgh's HBOS and Royal Bank of Scotland are the UK's two biggest losers. Outside the top ten, the high global ranking of Brussels (11th) and Washington, DC (13th) neither political world city is renowned for its banks internationally will be adjusted downwards as Fortis in Belgium disappears and Freddie Mac and Fanny Mae get downsized after their nationalization." (Taylor et al., 2009a: 10)

Of interest is now, what happens to these rankings if we study the data over several years. Is there really any transition? Is there really any transition during and after the global financial crisis of late 2007 and 2008, as Taylor et al. ask for:

"Will the demise of neoliberal globalization bring in its wake a lessening of economic globalization overall? Or will a new economic globalization be created in which the discarding of neoliberalism leads to less geographical concentration of command and control?" (Taylor et al., 2009a: 10)

5.2 Global cities before and after the global financial crisis

To answer the questions raised above, we need to have a look into the work of Ernest Ka Shing Lee, Simon Xiaobin Zhao and Yang Xie. They worked on these questions and tried to find out if there is a transition in the global space-economy before and after 2008. Based on the work of Taylor et al., yet they used a 5-year dataset of the Fortune Global 500 list to cover the years of 2005-2008. The years 2005-2006 outline the situation before the crisis, the year 2007 the year of "brewing and outset stage" (Lee et al., 2012: 335) and the year 2008 marks the post-outbreak situation. Here they also use the term command-and-control-cities for the terms world cities or global cities. Further, they use it for "cities as command and control centres and basing points". (Lee et al., 2012: 335) What is very important for the term "cities as command and control centres", that in this perspective these cities function as reference points of the global flows of capital. And this command and control functions which these cities possess, act "as a geo-economic indicator to visualize the adverse impact of global economic fluctuation". (Lee et al., 2012: 335)

5.2.1 Survey on the basis of the FCI

As mentioned above, they use the same methods as Taylor et al. but with a different data source. The data includes 100 financial service firms like banks, insurances and other multinational corporations which are in the Fortune Global 500 list. The resulting ranking shows the measured importance of a city as a command and control centre. The FDI is computed in the same way and with the same scores as Taylor et al. did. The results of Lee's et al. computing are shown in Tab.5. (Lee et al., 2012: 335-336)

First, a result that is not shown in Tab.5, is that the "financial service sector keeps expanding" (Lee et al., 2012: 336) over the years 2005-2009. The number of financial service firms raised from 108 firms to 119 but also 15 firms fell out the Fortune Global 500. That implies an impact of the "global credit crunch" on some financial service firms, so that their corporate size shrank. (Lee et al., 2012: 336)

According to Lees et al. computing, New York, Paris and London played the dominant role as the leading global command and control centres before the crisis. But New York lost the leading role in 2007 and Paris gained the place on the top. In terms of financial command and control functions, Western Europe was the significant region in the world.

But the tripod of New York, London and Paris did not remain after the crisis. In the year 2009 Tokyo and Beijing displaced London. So, London must be affected by the crisis, as it lost his peak at a score of 97.30 in 2006 (before the crisis) and had just a score of 78.57 in 2008 (after the crisis). This is a result of the revenue loss of banks, insurances and other financial firms which have their headquarters in London. Whereby, Tokyo stood on the same step like New York with a score of 86.11 in 2009, as Tokyo gained significantly command and control functions after the crisis in 2009. Very interesting is the rise of Beijing. Even while other cities lost control and command functions during the crisis, Beijing gained on and on command and control functions as we look at Tab.5. In 2005 Beijing stood on a rank of 13, in 2006 on a rank of 11, in 2007 on a rank of 12 but then in 2008 on a rank of 7 and in 2009 on a rank of 4. Beijing has still not comparable command and control functions in 2009 as we have a look at the score of 61.11 and at the score of Paris, New York or Tokyo. This narrowing between Beijing and London and New York, is a result by "the massive injection of capital support from the Chinese Government" (Lee et al., 2012: 336) and that State-Owned Enterprises in China are rather able to absorb the impacts of the global financial crisis than private enterprises in the capitalistic systems. (Lee et al., 2012: 336)

So, one could interpret this table as a shifting of command and control centres to the Asian region because North American and European cities suffered more from the global financial crisis than cities in Asia. Tokyo and Beijing should be the evidences here. (Lee et al., 2012: 336)

Besides the top command and control cities, there are other winners and losers. We can identify Zurich as a losing city after the crisis. First, Zurich was very stable on rank 5. But in the post-outbreak phase in 2008 Zurich felt back on rank 6 and in 2009 on rank 10. Like London, that is a distinct downfall. And in the years before and during the crisis Zurich lost continuous command and control functions with a continuous falling FCI score. The reason for this downfall is the shrinkage in revenue of five financial service firms which are headquartered in Zurich, UBS Credit Suisse, Zurich Financial Services, Swiss Reinsurance and Swiss Life. And also negatively impacted by the crisis was Brussels with its downfall from its rank 4 in 2008 to rank 9 in 2009. (Lee et al., 2012: 337)

Winnings cities in Europe were Munich and Edinburgh. In 2008 these cities ranked on 9 and 10, and in 2009 they ranked on 6 and 9. In Munich Allianz gained revenue and salvaged its balance sheet with the sale of Dresdner Bank to Commerzbank. The upward movement of Edinburgh could be explained due to smaller economic damage of

Edinburgh's HBOS and the Royal Bank of Scotland than for example financial service firms in Zurich, Brussels or Frankfurt. (Lee et al., 2012: 337)

Rank	City	FCI (2005)	City	FCI (2006)	City	FCI (2007)	City	FCI (2008)	City	FCI (2009)
1	New York	100.0	New York	100.0	Paris	100.0	Paris	100.0	Paris	100.0
2	Paris	91.89	Paris	100.0	New York	83.33	London	78.57	New York	86.11
3	London	86.49	London	97.30	London	80.95	New York	73.81	Tokyo	86.11
4	Tokyo	72.97	Tokyo	62.16	Tokyo	50.00	Brussels	47.62	Beijing	61.11
5	Zurich	56.76	Zurich	54.05	Zurich	50.00	Tokyo	47.62	London	55.56
6	Munich	54.05	Brussels	43.24	Brussels	38.10	Zurich	42.86	Munich	47.22
7	Amsterdam	40.54	Munich	43.24	Amsterdam	35.71	Beijing	35.71	Edinburgh	36.11
8	Brussels	35.14	Amsterdam	40.54	Munich	33.33	Frankfurt	33.33	Frankfurt	36.11
9	Osaka	32.43	Edinburgh	37.84	Charlotte	28.57	Munich	33.33	Brussels	33.33
10	Edinburgh	29.73	Frankfurt	29.73	Edinburgh	28.57	Edinburgh	30.95	Zurich	30.56
11	Frankfurt	29.73	Beijing	27.03	Frankfurt	28.57	Amsterdam	23.81	Amsterdam	27.78
12	Fairfield	27.03	Charlotte	27.03	Beijing	26.19	Charlotte	23.81	Fairfield	27.78
13	Beijing	24.32	Fairfield	27.03	Fairfield	23.81	Fairfield	23.81	Madrid	25.00
14	Charlotte	24.32	Osaka	21.62	Toronto	21.43	Toronto	23.81	Charlotte	22.22
15	Omaha	21.62	Toronto	21.62	Osaka	16.67	Milan	14.29	Osaka	22.22
16	Trieste	21.62	Trieste	21.62	Omaha	14.29	Omaha	14.29	Toronto	22.22
17	Bloomington	16.22	Omaha	16.22	Trieste	14.29	Trieste	14.29	Milan	19.44
18	Toronto	16.22	Bloomington	13.51	Bloomington	11.90	Bloomington	11.90	Omaha	16.67
19	Madrid	10.81	Madrid	10.81	Madrid	11.90	Madrid	11.90	Trieste	16.67
20	Mclean	10.81	Mclean	10.81	Mclean	11.90	Osaka	11.90	Bilbao	13.89

Tab. 5: global city ranking according to the FCI 2005-2009, source: Lee et al., 2012: 337

5.2.2 Survey on the basis of the BCI

To study the impact not only on the financial sector but also on other industries, Lee et al. used the Business Command Index and the Fortune Global 500 as data source. Here, one could interpret the data to conceive the impact before and after the financial crisis "on the command and control functions of cities measured by the largest 500 headquarters of multinational corporations". (Lee et al., 2012: 338)

Computing the BCI and scoring happens in the same way like Taylor et al. 2009 did. The results are shown in Tab.6.

Obviously, the spatial distribution is clearly across all over the world. But North America, Europe and North East Asian are playing the dominant "role commanding the business operations globally" in the computed five years 2005-2009. (Lee et al., 2012: 339)

Tokyo is the stead leading command and control city followed by Paris, New York, London and Beijing. In the phase of recession, in 2009, New York and London, as "traditional global business command nodes" (Lee et al., 2012: 338), each lost a rank and felt down on rank 4 and 5, while Beijing, as a city of a developing economy, climbed upon rank 3. As cities like Paris, London and New York are still under the top five command and control centres, one could suggest that firms in the financial sector absorbed the impact of the global crisis so that firms of other sectors are less shattered by the global financial crisis. But that does not mean that firms of other sectors were not impacted by the crisis. That can be attested by the BCI values in Tab.6 of London, Paris and New York the years 2007-2009 where all three cities had increasing BCI values but decreasing values in the recession phase 2009. In contrast, Tokyo without losses and Beijing with a constantly increasing BCI value, even in the phase of recession in 2009. (Lee et al., 2012: 338)

Rank	City	BCI (2005)	City	BCI (2006)	City	BCI (2007)	City	BCI (2008)	City	BCI (2009)
1	Tokyo	100.0	Tokyo	100.0	Tokyo	100.0	Tokyo	100.0	Tokyo	100.0
2	Paris	63.50	Paris	67.46	Paris	68.04	Paris	72.04	Paris	60.28
3	New York	54.87	New York	60.29	New York	65.46	New York	60.75	Beijing	51.64
4	London	43.36	London	50.24	London	52.58	London	52.69	New York	38.79
5	Beijing	23.01	Madrid	35.17	Beijing	36.08	Beijing	44.09	London	32.24
6	Munich	21.24	Beijing	27.75	Seoul	24.23	Seoul	30.65	Seoul	22.43
7	Seoul	18.58	Seoul	21.53	Munich	21.13	Munich	23.12	Munich	19.63
8	Zurich	16.81	Munich	20.57	Zurich	18.56	Madrid	18.82	Madrid	17.76
9	Houston	14.16	Zurich	17.70	Madrid	17.53	Zurich	18.28	Moscow	14.95
10	Dusseldorf	12.83	Amsterdam	16.03	Amsterdam	16.75	Osaka	15.05	Houston	14.49
11	Osaka	12.83	Houston	15.31	Houston	16.49	Brussels	14.52	Dusseldorf	14.02
12	Atlanta	12.39	Dusseldorf	13.40	Brussels	13.92	Dusseldorf	14.52	Osaka	12.62
13	Amsterdam	11.50	Brussels	12.44	Dusseldorf	13.40	Houston	14.52	Rome	11.21
14	Madrid	11.50	Rome	11.96	Rome	12.89	Amsterdam	12.90	Stuttgart	11.21
15	Rome	11.06	Atlanta	11.48	Atlanta	11.34	Stuttgart	12.37	Zurich	11.21
16	Bonn	8.41	Edinburgh	11.48	Stuttgart	11.34	Moscow	11.83	Amsterdam	10.28
17	Brussels	8.41	Osaka	11.48	Edinburgh	10.82	San Antonio	11.83	Atlanta	8.88
18	Cincinnati	8.41	Cincinnati	9.57	Cincinnati	10.31	Edinburgh	11.29	Cincinnati	8.88
19	Courbevoie	8.41	Courbevoie	9.09	Courbevoie	9.79	Frankfurt	11.29	Courbevoie	8.88
20	The Hague	8.41	Moscow	9.09	Frankfurt	9.79	Atlanta	10.75	Frankfurt	8.88

Tab. 6: global city ranking according to the BCI 2005-2009, source: Lee et al., 2012: 339

6. Shifting of leading global cities? – A conclusion

Scientists agree that there are world cities, global cities or command-and-control centres. This arises out of the intensifying processes of globalization. The idea is that there are centres which play a dominant role in the global economic interweaving. The whole debate about global cities has an economic ductus. In some definitions, global cities are centres of politics, economics, communication, culture and science. But the focus is obviously on the role of global cities as anchorage points and control centres within the organisation of the global economy.

With the intensifying processes of globalization, there must be a rise of a global urban network and a demand for cities as centres of commanding and organizing the global economy in order to function. Friedmann calls world cities as the geographical emanations of such commanding centres and that the modern world-system is defined by its world cities which are the nodes in networks of power and dominance.

And as Sassen mentions, global cities have a functional centrality in the global economy and that power is essential to organize world economy. Global cities are the centres of dominance and power or centres of global control.

Friedmann suggests that there is a spatial inequality in this kind of cities as a result of division of labour. That is a feature of the fordistic city. And now, the global financial crisis comes into play. As Nölke stated, that there would no actual crisis without financialization and such a crisis could never have broken out under Fordism. With neoliberal politics promoted mainly in the USA and Great Britain by Reagan and Thatcher, financialization appeared. So, the financial sector got more and more important and that means we must have a look at the development of the global cities as financial centres. With the financialization, the highly financialized economic systems, especially the USA and Great Britain are increasingly crisis-prone. Under neoliberalism, we have the neoliberal city which acts like a corporation. And so, they are also highly crisis-prone.

Nölke mentioned that the global economic crisis highly impacted the Anglo-Saxon heartland, and less continental Europe and relatively less emergent economies like India, South Africa or Latin America. As shown in Tab.2, where the crisis had its biggest impacts namely on corporations which are headquartered in the USA and Great Britain.

And this coincides with the studies of Lee et al., that leading cities with headquarters of financial service firms (New York and London) lost command-and-control functions after the crisis. While these cities lost command-and-control functions, Asian cities like Tokyo or Beijing were not or less impacted by different reasons. Tokyo has a stable ranking and Beijing is an upcoming global city according the studies of Lee et al.

When we interpret the results of Lee et al. one can say that there is a shifting of leading global cities, or transition, from the Anglo-Saxon region into the Asian region and global cities surely suffered from the global economic crisis or of the following recession phase. But to make a clear statement about a persisting shifting into Asia, there must be more analysis of the years after 2009.

And that is a problem here about the given literature. There are a small number of scientists who analysed the impacts of the global financial crisis on global cities and who go further into the question, if there is a shifting of leading global cities from the Anglo-Saxon region into the Asian region. On the basis of given literature, one could reply the question positively. To speculate a little, with the upcoming emergent economies like the BRICS-states and their increasing importance within the world economy, one could suggest that also some cities in these countries will rise and gain more and more importance within the global cities network. But there need not to be a shifting. Cities in, e.g. the BRICS-states could just close the lines to cities like Tokyo, Paris, London or New York and not relieve these cities. So, there need to be further investigations to give a clear statement about these questions.

For further investigations, one needs comparable and meaningful methods. There must be some disagreement and ambiguity about the methods. To clarify, if there is effectively a transition of leading global cities towards Asia and if this transition is crisis-made, on has to decide between different methods. One could use the method used by Bronger (2011). As he considers the global city concept as functional concept, he uses eight indicators (headquarters of the biggest TNC's by quantity and volume, biggest banks by assets, biggest stock exchanges by volume, important airports by passengers and cargo volume, important seaports by handling, headquarters of important institutions) to find out what global cities are and to create a ranking to show the importance of these cities within the world economy. But this just creates a vertical ranking. One could use also the question in what ways the global financial crisis impacted global cities. With this method you could detect if some cities lost importance and other cities take the lead.

But what could be more interesting for geographers is to study the horizontal ranking. Therefore, Taylor created the interlocking network model to detect in what ways global cities are connected into the global city network and in what relations each city stands to another city. Hence, you get a value that indicates the network connectivity of a city concerning the network. Then one could find out if some cities lost network connectivity or if some cities gained network connectivity after the crisis, and if there is a shifting of previously well connected cities in the Anglo-Saxon and European region towards the cities in Asia after the crisis.

Instead of using that method, Taylor et al. use another method to find out about a conceivably transition. And then Lee et al. use the same method, but with another data source. So, to study the network connectivity is a more detailed and more meaningful method for the raised questions above. But this method is much expensive because there is a lot of data which one has to compute. For a short overview and to get a quick result, the method with computing the FCI and BCI is a useful one.

At the end, scientists even do not go along with the definitions about global cities. So it is hard to decide what method to use, first to identify the global cities and then to examine the impacts of the global financial crisis on global cities and furthermore if there is any shifting of the importance of leading global cities towards Asia.

7. References

BRONGER, D. (2011): Megastädte – Global Cities HEUTE: Das Zeitalter Asiens? Berlin

DERUDDER, B. (2012): International Handbook of GLobalization and World Cities. Cheltenham

LEE et al. (2012): Command and Control Cities in Global Space-economy before and after 2008 Geo-economic Transition. In: Chin. Geogra. Sci. 2012, vol. 22, no. 3 pages 334–342

NÖLKE, A. (2009): Finanzkrise, Finanzialisierung und Vergleichende Kapitalismusforschung. In: Zeitschrift für internationale Beziehungen, vol.16, no. 1, pages 123-139

TAYLOR, P.J. et al. (2009): The way we were: command-and-control centres in the global space-economy on the eve of the 2008 geo-economic transition. In: Environment and Planning, vol. 41, pages 7-12